IMAGES
of America

LOUISVILLE
TELEVISION

D1637529

Ed Kallay of WAVE poses on the set of his children's show, *Funny Flickers*, in the 1950s. Also known as *The Magic Forest,* the show ran from 1953 to 1965.

IMAGES
of America

LOUISVILLE
TELEVISION

David Inman

ARCADIA
PUBLISHING

Published by Arcadia Publishing
Charleston, South Carolina

Printed in the United States of America

Library of Congress Control Number: 2010925479

For all general information, please contact Arcadia Publishing:
Telephone 843-853-2070
Fax 843-853-0044
E-mail sales@arcadiapublishing.com
For customer service and orders:
Toll-Free 1-888-313-2665

Visit us on the Internet at www.arcadiapublishing.com

To my mom and dad, Marilyn and David Inman

CONTENTS

ACKNOWLEDGMENTS

When I first began working at the *Louisville Times* in 1981, I would get the strangest feeling of déjà vu every time I walked past a freight elevator located just outside the newsroom. It took me a few weeks to realize that my familiarity came from the days when WHAS-TV was in the same building as the newspaper, and that the freight elevator was the same one I had been herded into with a couple dozen other kids to attend a broadcast of *T-Bar-V Ranch* back in 1962 or 1963. I hope this book of images and memories gives you the same feeling of nostalgia.

In compiling it, I had help from Melanie Bullard, Wally and Barbara Dempsey, Ray Foushee, Jeff Hendricks, Eileen Douglas Israel, Ronald Lewis, David R. Lutman, Jayne McClew, John Quincy, Chris Shaw, and Steve York.

At Arcadia Publishing, editor Amy Perryman has been the essence of professionalism and promptness in response to any questions or concerns. It has been a pleasure working with her.

I also have to thank those who have been so generous with photographs and memories in the past: Randy Atcher, Bob Bowman, Bob Fulbright, Sam Gifford, Louise and Ryan Halloran, David Jones, Mike Kallay, Bob Kay, Phyllis Knight, Sleepy Marlin, Milton Metz, Ray Moran, Monnie Walton Parker, Bob Pilkington, Jerry Rice, Kaelin Kallay Rybak, Julie Shaw, and Ray Shelton.

As a lifelong resident of Louisville and a lifelong consumer of local television, it has been an honor for me to meet these people, hear their stories, and relate their experiences to readers who were there at the beginning as well as those who have come along in the ensuing years. Any sins of omission or commission are the fault of the author and no one else.

Unless otherwise indicated, all photographs are from the author's collection.

INTRODUCTION

Once upon a time, back when television was a toddler, it ate local shows for breakfast, lunch, and dinner. There were no reruns; they hadn't been invented yet. There were no syndicated shows like *Oprah* or *Live With Regis and Kelly*. There were network shows, and there were local shows. Local shows were different in every city—as unique as the city itself.

In Louisville, if a child was a Boy or Girl Scout, the scout and the troop might have appeared on *Hayloft Hoedown*, waving at the camera and telling riddles to the show's comedy relief, "Cactus" Tom Brooks. A teenager might have been on *Teen Beat* or *Hi Varieties*, doing the twist or playing "Too Fat Polka" on an accordion. An adult might have been interviewed on *Small Talk* or *The Morning Show* or *Omelet*. And, of course, on his or her birthday, someone might have made an appearance on *T-Bar-V Ranch*. Even those who did not make a cameo but grew up in Louisville in the 1950s and 1960s know the opening and closing theme:

> Howdy, howdy, boys and girls
> It's T-Bar-V Ranch time
> We're glad to see you all today
> And hope you're feeling fine
> We'll sing and dance and have a show
> And birthday parties too
> It's T-Bar-V Ranch time!
>
> Brush your teeth each morning,
> get lots of sleep at night
> Mind your mom and daddy,
> 'cause they know what is right.
> Lots of exercise each day
> and eat up all your food
> And always wear a great big smile
> that makes you look so good
> Be sure to look both left and right
> before you cross the street
> And be with us tomorrow at nine
> when it's T-Bar-V Ranch time!

Local television made stars of the people who hosted or performed on these shows, and it gave viewers a chance to be stars in their neighborhoods, if only for a day or two. Long before the term was coined, local TV was interactive, offering access to people in the community who had a good story to tell, a song to sing, or a birthday to celebrate. In the mid-1950s, African

Americans were still segregated in Louisville movie theaters and at lunch counters, but they could appear with Randy and Cactus on *T-Bar-V Ranch* or audition for the king or queen of the "WHAS Crusade for Children" on *Hi Varieties*. Muhammad Ali's boxing career began on WAVE's *Tomorrow's Champions*.

Local television stations still exist, of course, but except for news, local programming really does not exist any more. For a long time now, it has been cheaper to run syndicated programming, which is why today people see the same shows whether they are in Los Angeles, Lubbock, or Louisville. To add insult to injury, most local television was produced in the days before cheap, accessible videotape, so those regional shows remain only in memories and photographs. Here readers will find both, as well as a look at Louisville television that encompasses the Internet age as well as the golden age.

Television has come a long way since its toddler days. Whether that is good or bad has occupied many a debate. The intent of this book is to show the growth process, at least as it happened here.

One

1948–1961

ON THE AIR

Louisville television began in a studio on Broadway with singers, dancers, cornpone comics, and Junior, a pop-eyed dummy in a houndstooth jacket. The initial program on WAVE-TV aired the day before Thanksgiving, Wednesday, November 24, 1948.

Louisville became the 22nd American city to begin regularly scheduled television programs, and WAVE was the first television station in Kentucky. Then on the frequency of channel 5, WAVE signed on the air with a lineup including emcee Burt Blackwell, comic O. B. Carpenter, bandleader Clayton "Pappy" McMichen and his Georgia Wildcats, the Libby Starks square dancers, and ventriloquist Norma Jarboe. Jarboe and her dummy Junior became WAVE's first stars. The success of her appearance on the station's debut show led to her own series, *Junior's Club*, cohosted by Ed Kallay, who would soon host his own long-running children's program, *Funny Flickers*, on WAVE. Junior, meanwhile, soon became the hardest-working dummy in show business. His popularity led to WAVE adding *Junior's Sketchbook*, *Junior's Movie*, and *Junior's Pet Show* to the schedule.

At the time, there were only about 2,000 televisions in town, many of them installed by a team of experts—either from the store where the set was purchased, the factory that made the set, or from WAVE itself—to ensure the best reception.

Because the coaxial cable did not yet stop in Louisville, there were no live network shows. WAVE relied on kinescopes, filmed versions of live shows usually several days old and usually delivered by bus. Since WAVE was the only station in town, it carried programs from all the national networks, including *Arthur Godfrey and His Friends* and Ed Sullivan's *Toast of the Town* from CBS, *Kukla, Fran and Ollie* and Perry Como's *Chesterfield Supper Club* from NBC, and *Cavalcade of Stars* from the DuMont network.

A little over a year later—on March 27, 1950, to be exact—WHAS-TV signed on the air. One of its first shows, *T-Bar-V Ranch*, would stick around for the next 20 years.

During the inaugural broadcast on WAVE-TV in 1948, director and emcee Burt Blackwell did a routine with comic O. B. Carpenter. Seen at left are other performers—ventriloquist Norma Jarboe and the Libby Starks square dancers. At right is bandleader Clayton "Pappy" McMichen. The opening show lineup also included Mary Ann Miller, Bob Reed, Bea Davidson, and students from the Lilias Courtney Dance School.

In the early days of television, announcers did double or even triple duty. Livingston Gilbert of WAVE did everything from live commercials to introducing a movie each day at 4:00 p.m. He is best known, of course, as a WAVE television and radio news anchor, a position he held from 1941 until 1980.

A video version of Ryan Halloran's radio show, *Dialing for Discs*, was one of the first programs on WAVE-TV. One day, a woman whose grown daughter was recuperating at a nearby tuberculosis sanitarium called the program to say that her daughter had recently given birth but could not see the baby because she was hospitalized. The baby was brought into the station and put on television so the mother could see it.

In the 1960s and 1970s, Rodney Ford was the sedate editorial spokesman for WAVE. But in the late 1940s and early 1950s, he played backwoods comic character Burley Birchbark on WAVE radio and TV.

Rodney Ford (below, seated at left) is Burley Birchbark on a musical show in the early 1950s. Standing with his foot on the barrel is WAVE announcer Bob Kay.

Locally produced programs included newscasts, kiddie shows, musical programs, and even game shows. Rosemary Reddens (center) and Bob Kay were the cohosts of the WAVE game show *Pop the Question*, which ran from 1951 to 1954.

In 1953, WAVE moved from channel 5 to channel 3 in order to expand its broadcast reach. At the broadcast commemorating the changeover was WAVE staffer Bill Gladden, shown above.

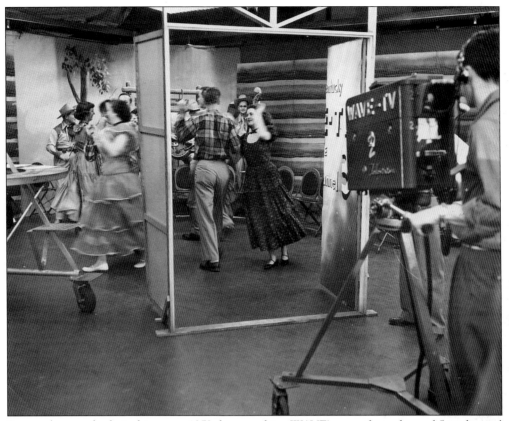

Square dancers do their thing on a 1953 show marking WAVE's move from channel 5 to channel 3 for a wider reception range.

Ed Kallay introduces WAVE's wider broadcast area as it switches from channel 5 to channel 3 in 1953.

News coverage was primitive in the days of early TV. Shown above is an example of how WAVE covered news stories: by shooting a close-up of a Polaroid photograph illustrating the event.

By the late 1950s, newscasts were a little more sophisticated. Here Ryan Halloran does a telecast with visual aids on WAVE.

Ed Kallay stands on the *Funny Flickers* set in the late 1950s.

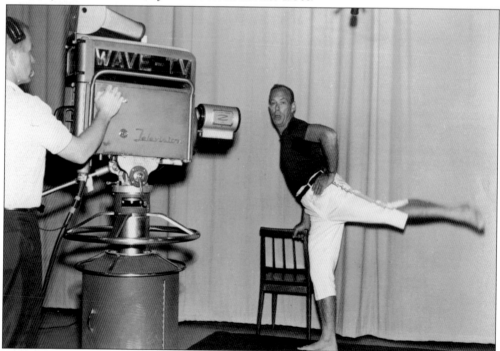

Ed Kallay was not just a kiddie-show host. He was an announcer as well as the WAVE sports director, sportscaster, and play-by-play broadcaster. For a while in the 1950s, Kallay even hosted an exercise show.

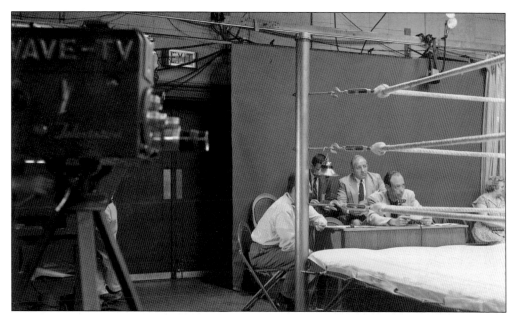

Ed Kallay (far right) also anchored WAVE's *Tomorrows Champions,* which featured young boxers from the area and helped give Cassius Clay (Muhammad Ali) his start. Here Kallay moderates from ringside in the mid-1950s.

In the early 1950s, Ed Kallay does a sportscast from a crowded WAVE set.

Ed Kallay's sidekicks on *Funny Flickers* were two puppets—Sylvester the Duck and Tom Foolery. Here Kallay poses with Sylvester. Two of Kallay's catchphrases on the show were "See you 'round like a donut!" and "So long, like a hot dog!" On the stump in front of Kallay are a copy of his script and a package of Klarer hot dogs, which was one of the show's sponsors.

PROGRAMS IN THE "BEST" TRADITION

Every night at 10:30 Ray Shelton and Kentuckiana's exclusive local newsreel. Sponsored by Greater Louisville First Federal Savings and Loan Assoc.

Every Friday at noon Lee Jordan, Allan Blankenbaker help local women have fun and prizes. Sponsored by Triangle Food Stores.

WHAS-TV *Channel 9*

POWER   POPULARITY

When it went on the air in 1950, WHAS-TV was on the frequency of channel 9. It would not change to channel 11 until 1953.

The first newscaster on WHAS-TV was Pete French, a longtime announcer with WHAS radio.

From left to right, "Cactus" Tom Brooks and Randy Atcher perform on *T-Bar-V Ranch*. Atcher had been a singer on WKLO radio, and when WHAS made him an offer, he told WLKO management he would stay if they paid him what they were paying announcer Foster Brooks. (Atcher knew they were paying Brooks's salary as well as his alimony.) WKLO refused, and Atcher went to WHAS.

From left to right "Cactus" Tom Brooks, Randy Atcher, and Carl "Tiny" Thomale are on the set in the early 1950s. Before he played Cactus, Tom Brooks was an announcer whose career had been iffy because of nerves and stammering. But when he took out his false teeth to play a cowboy bumpkin on the radio, Cactus was born. Brooks's trademark was a loud "How-DEEEEE!"

Greater Louisville First Federal Savings and Loan was an early sponsor of *T-Bar-V Ranch* and set up a "Savings Post" in its downtown office where children could open bank accounts. By 1956, over 20,000 youngsters had saved over $1 million. In this 1954 shot, Randy Atcher poses with Ronald Lewis. (Ronald Lewis.)

Randy Atcher (left) and "Cactus" Tom Brooks pose with a young customer at the "Savings Post."

On another day at the "Savings Post," Randy Atcher sings a song while a little girl peruses a brochure.

WHAS announcer Ray Shelton sings the praises of Folger's Coffee. Shelton was with WHAS from 1950 until 1975, leaving to join Greater Louisville First Federal Savings and Loan. He did radio and TV spots for the bank from 1950 through the mid-1990s.

WAVE's Bob Kay, dressed as a deliveryman for Donaldson Bakery, was the host of *Donaldson Play Club,* which ran from 1953 to 1955. One day on this show, instructed to sell the sponsor's product as "the best in bread," Kay accidentally said "the breast in bed."

Foster Brooks and Ryan Halloran cohost the game show *Go for Cash* on WAVE in 1954.

When television was new, technical difficulties were common. This is what WAVE viewers saw when there were problems.

Burt Blackwell (standing, in necktie) directs a WAVE program in the early 1950s.

On a warm summer day in the early 1950s, WHAS-TV personality Bud Abbott (in sunglasses) decided to do his show from the roof of The Courier-Journal building, accompanied (in more ways than one) by Tiny Thomale. Abbott's comedy-interview show aired on WHAS from 1950 to 1956, when he left the station to join Radio Free Europe.

As part of his WHAS show, Bud Abbott interviews an indifferent Spike Jones and a not-at-all-indifferent Sally Rand. Jones was a comic/bandleader, and Rand was a well-known exotic dancer, hence the balloons in the background. Both photographs are from the early 1950s.

A live broadcast of the WHAS show *Walton Calling* in the early 1950s is hosted by Jim Walton (center, in light suit) and brought to you by Will Sales. A stunt "staging area" is at left. *Walton Calling* was the TV version of Walton's *Coffee Call* radio show.

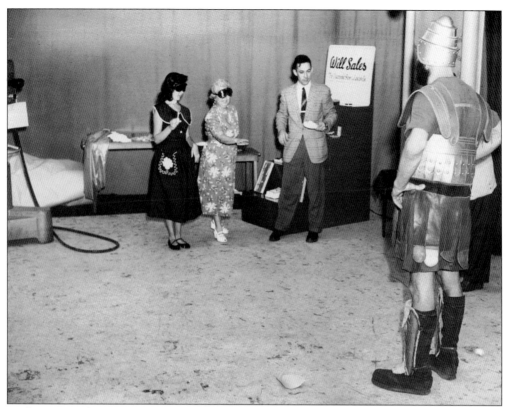

Audience members perform a stunt involving blindfolds and a suit of armor on Jim Walton's program called *Walton Calling* on WHAS in the early 1950s.

Created in the early 1950s by WHAS artist Allen Blankenbaker and producer Bill Loader, FISBIE became the station logo and mascot. FISBIE stood for "Foremost in Service, Best in Entertainment."

On the second anniversary of his *Coffee Call* program in 1950 on WHAS, Jim Walton had some help enjoying the cake.

WAVE tried to counteract the success of WHAS's *T-Bar-V Ranch* with *The Old Sheriff*, which ran from 1950 until 1953. The title role was played by Foster Brooks, brother of WHAS's Tom Brooks, Cactus on *T-Bar-V Ranch*. Foster Brooks would leave Louisville in the late 1950s and become a popular comedian on 1970s talk and variety shows, famous for his "drunk act."

Foster Brooks, with glasses and mustache, poses with local personality Boyd Bennett on the set of WAVE-TV's short-lived *Boyd Bennett and His Space Buddies* in 1952.

WHAS personality Phyllis Knight, a pioneer in women's broadcasting, interviewed the famous and not-so-famous on her program *Small Talk* from 1956 to 1969. In this shot from the late 1950s, she talks to actor and future president Ronald Reagan; he was a spokesman for General Electric and made regular visits to Louisville's Appliance Park.

Phyllis Knight was also a trailblazer in heightening awareness of breast cancer. She helped promote special cancer films at local movie houses, including at the Rialto Theatre in downtown Louisville in 1958.

WHAS's Phyllis Knight and Sam Gifford, married in real life, were doing a show together in the mid-1950s.

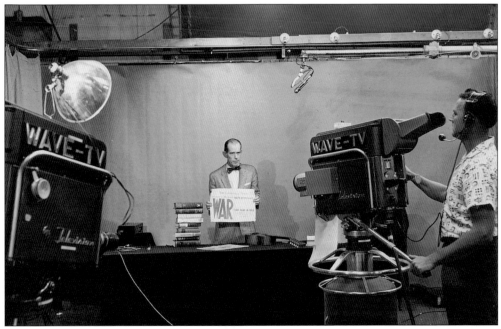

In 1958, educational broadcasts began over station WFPK-TV (later WKPC). Broadcasts originated from the WAVE studios and featured history instructor John Dickey.

As WHAS weatherman for most of the 1950s and 1960s, Milton Metz had his trademarks. He would always give the temperature in Caribou, Maine, and the forecast would appear on a "Magic Weather Writer," some camera trickery devised by Metz's wife, Miriam. Metz would also give the highest and lowest temperature in the country. When his sponsor was an appliance dealer, iceboxes and stoves were used for the temperature extremes (right), and when the sponsor was Kool cigarettes, the Kool penguin was featured (below).

One of WAVE's first musical shows was *The Pee Wee King Show,* which ran from 1948 through 1954. Pictured are King (playing the accordion) and his band, the Golden West Cowboys, including vocalist Redd Stewart (the fiddle player at the rear of the photograph, near the door).

Pee Wee King (left, on the set) was shown with Bob Kay on *The Pee Wee King Show* in the late 1950s. The program was sponsored, obviously, by Oertel's '92 Beer.

Pee Wee King and his band, including vocalist Redd Stewart (first row, far right) and announcer Bob Kay (first row, far left), are on the set of their WAVE show in the late 1950s.

The Motet Singers from St. Paul United Methodist Church appear on the WHAS show *Hi Varieties* in the late 1950s.

Shorty Chesser, Randy Atcher, and Shirley Cardinal, shown from left to right, sing a number during the 1953 *Hayloft Hoedown* Christmas show.

The entire *Hayloft Hoedown* gang is on the set at WHAS in 1953. They are, from left to right, Sleepy Marlin, Bea and Mary House, Shorty Chesser, Tiny Thomale, Bernie Smith, George and Janie Workman, and Randy Atcher. The show was carried in prime time and attracted audiences as large as the CBS shows that surrounded it on the schedule, like *Our Miss Brooks* and *The Twilight Zone*.

Singer Judy Marshall takes the stage on WHAS's *Hayloft Hoedown* in the late 1950s.

Making music during a late 1950s broadcast of *Hayloft Hoedown* are Tiny Thomale (left), George Workman (center), and Bernie Smith.

Randy Atcher (far right) is backed by, from left to right, Tiny Thomale, Shorty Chesser, George Workman, and Bernie Smith on *Hayloft Hoedown* in the mid-1950s.

The Sharpe twins, Janice (left) and Janette, were regulars on *Hayloft Hoedown* in the early 1950s.

In the 1950s, the Hayloft Hoedowners, the show's square dancing group, performed on *Hayloft Hoedown*.

Shorty Chesser and Judy Marshall sang a duet on *Hayloft Hoedown* in the late 1950s.

Tiny Thomale (left) and "Cactus" Tom Brooks do a routine on *Hayloft Hoedown* in 1957. The show's audience often included local Boy and Girl Scout troops who would share jokes with Cactus.

In 1952 and 1953, WHAS held cerebral palsy telethons that were the predecessors of the "WHAS Crusade for Children." Mel Torme (left) came to town for the 1953 show. Here he solemnly listens to information from emcee Jim Walton. Legend has it that Torme was annoyed because local favorite Randy Atcher got more applause than he did.

At the 1959 "WHAS Crusade for Children," emcee Jim Walton (center) soaked his feet along with singer Tommy Leonetti (left) and actor Lee Marvin (right). The telethon is one of the longest-running in the country, still airing each year. Walton was emcee of the crusade from 1953 until 1979. When he died in 1985, his funeral procession included volunteer firefighters from all over Kentuckiana.

In the 1950s, WAVE-TV held several "Bids for Kids" telethons. In 1959, celebrities included, from left to right, country comic Minnie Pearl (out of costume), game show emcee Hal March (*The $64,000 Question*), and singer Snooky Lanson (*Your Hit Parade*). They are overseen by Ed Kallay.

Announcers who were with WAVE-TV when it signed on in 1948 celebrate the station's 10-year anniversary in 1958. From left to right are Bill Gladden, Ed Kallay, Ryan Halloran, Livingston Gilbert, and Bob Kay. All except Gladden would stay with the station until they retired.

Two

1961–1971
BOOM AND BUST

In 1961, Louisville received its third television station when WLKY, an ABC affiliate, went on the air. It was a UHF station, channel 32, so reception at first was spotty. A year later, NBC affiliate WAVE began broadcasting local shows in color, beginning with its new daytime entry, *The Morning Show.* Around the same time, local stations began using videotape. This meant that Louisville-area kids could actually see themselves on *T-Bar-V Ranch,* because the shows were taped the day before.

As the decade began, local TV programs were as popular as ever. *Hayloft Hoedown* was still in prime time. As late as 1968, Randy Atcher and "Cactus" Tom Brooks of *T-Bar-V Ranch* were still icons. An article in the *Louisville Times* reported that after Cactus signed a child's arm at a personal appearance, the child refused to wash it off. A week later, the mother phoned Cactus, who urged the child to take a bath and promised to mail an autographed photograph.

But there were rumblings. There were now more options to replace local programming—reruns of popular shows or new syndicated shows. And they were cheap—much less expensive than producing local shows. On top of that, times were changing, and local shows were starting to appear somewhat quaint.

A vivid example happened in the spring of 1969, when WAVE began running NBC's hit show *Rowan and Martin's Laugh-In* on Saturday nights opposite *Hayloft Hoedown* on WHAS. Viewers could not have asked for a clearer contrast: it was square dancing versus go-go dancing, Randy and Cactus versus Rowan and Martin, gospel music versus Goldie Hawn. For the first time in 20 years, *Hayloft Hoedown* began fading in popularity.

Then in June 1970, one shoe dropped when *T-Bar-V Ranch* went off the air, abruptly cancelled after 20 years. There was no warning, no farewell show. It was replaced by reruns of *My Favorite Martian.* There were promises of work as announcers and a future at WHAS for Atcher and Brooks, but all they had left was *Hayloft Hoedown,* and it was cancelled in January 1971. That March the other shoe dropped when Atcher, 52, and Brooks, 61, were let go.

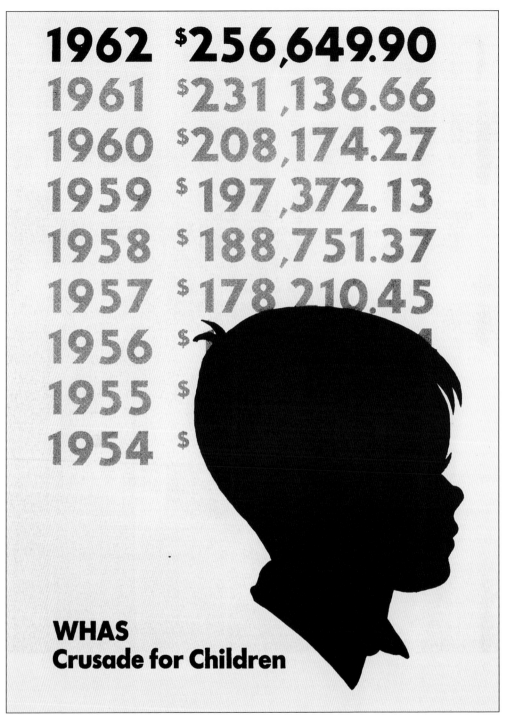

1962 $256,649.90
1961 $231,136.66
1960 $208,174.27
1959 $197,372.13
1958 $188,751.37
1957 $178,210.45
1956 $
1955 $
1954 $

**WHAS
Crusade for Children**

A 1962 brochure from the "WHAS Crusade for Children" demonstrates the telethon's rising totals year over year, a trend that continues today.

Phyllis Knight and Randy Atcher help kick off the 1961 "WHAS Crusade for Children."

The 1961 "WHAS Crusade for Children" takes place on the stage of Memorial Auditorium.

In 1961, as today, volunteer firefighters are the primary fund-raisers for the "WHAS Crusade for Children." Here Phyllis Knight interviews a group.

Phyllis Knight and Jim Walton pick out souvenir fire hats together for the 1965 "WHAS Crusade for Children."

WHAS-TV, Louisville, Ky.

5:30-6:30 p.m. Saturday - HI VARIETIES

Vocalists . . . ventriloquists; musicians . . . comedians . . . impressionists; dancers . . . drummers—"Hi Varieties" is a firmly established showcase program for teen-age talent of Kentucky and southern Indiana. It has been seen on WHAS-TV since 1950 when the station went on the air. TV Program Director Sam Gifford is master of ceremonies.

Dealing only with the teen-age group, the show presents all types of talent, confined usually to individual acts; however, it is not unusual to see 50 to 100 people in the cast, since high school bands and choral groups are frequent participants.

This popular medium of expression for talented youngsters who have a flair for showmanship has strong appeal for both teenagers and parents.

Several "Hi Varieties" regulars have become professional performers. Judy Marshall, a staff vocalist at WHAS for four years, was a "Hi Varieties" singer. Guitarist-singer Bobby Lewis now sings on WHAS-TV's "Hayloft Hoedown." WHAS vocalist Jo Ann Hale graduated to this same program. Two former WHAS staff vocalists started on "Hi Varieties."

A vocal motet, which presents various musical selections, and the "Hi Varieties" Dance Line, which provides lively rhythmic steps, appear on each program with emcee Sam Gifford.

The WHAS Crusade For Children conducts its annual King and Queen contest on this show, with a series of eliminations usually extending over some eight programs.

WHAS-TV *Fisbie* 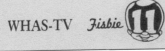 "Foremost In Service—Best In Entertainment"

Each "Crusade for Children" had a teenage king and queen who were winners of a talent show held on the WHAS program *Hi Varieties*. This sales sheet details the contest as well as the show's general format.

In 1962, *The Morning Show* premiered on WAVE. It offered a daily dose of talk, entertainment, and news. Ryan Halloran was a cohost.

Julie Shaw was Halloran's cohost on *The Morning Show*.

A fashion show on *The Morning Show* featured mid-1960s styles.

The Morning Show was the first local program to be broadcast in color.

Small Talk
WITH PHYLLIS KNIGHT

Small Talk is the feature section of FOCUS . . . WHAS-TV's nightly, 60-minute television journal from 6 to 7 p.m. In this swiftly moving, full hour report of the news and people making today's history, viewers are kept abreast of local, national and world news, sports and weather news, and important people visiting the area.

Winner of a McCall's "Golden Mike Award" as the "outstanding woman broadcaster in America," Phyllis is a skilled performer of 15 years before the microphone.

Her guest register regularly includes the leading "name" visitors to the city. Dmitri Shostakovich made his initial live television appearance in the U. S. on "Small Talk." Other programs have featured jazzmen Jonah Jones and Charlie Shavers; national columnist Inez Robb; entertainers Jimmy Dean and Lee Marvin; Abba Eben, Israeli Ambassador to the U. S. on the day that he resigned that diplomatic post; and many others.

"Small Talk" remotes, both live and video taped, have originated from Churchill Downs, the Kentucky State Fair, the nation's only floating Coast Guard Station and Bowman Air Field, as Phyllis became the first Kentucky woman to fly in a jet fighter plane.

Phyllis is briefed before a remote.

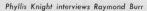
Phyllis Knight interviews Raymond Burr

Phyllis has played a vital role in many important community causes, including cancer-survey projects, the Louisville Fund and the spectacular WHAS "Crusade For Children."

Local and national sponsors have demonstrated their enthusiasm for this program and its friendly and sincere hostess. The audience which "Small Talk" reaches is the heart of the buying power of the Kentuckiana area.

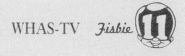

WHAS-TV *Jisbie* **11** "Foremost In Service—Best In Entertainment"

A 1962 sales sheet touts the virtues of Phyllis Knight and her WHAS program *Small Talk*. During the 1960s, the show explored topics such as sex education and race relations. For one series of reports, host Knight traveled to an Eastern Kentucky town to interview the first doctor in the area to prescribe birth control pills.

In the early 1960s, *Perry Mason* star Raymond Burr (above, left) visited Phyllis Knight on *Small Talk*.

Phyllis Knight talked to former first lady Eleanor Roosevelt in the 1950s.

Phyllis Knight is shown above doing a live spot at Churchill Downs in 1965.

In addition to cohosting *The Morning Show*, Julie Shaw was Miss Julie the Story Lady on WAVE's daily *Funny Flickers* kids show. Shaw wore an apron in which she kept storytelling props. She came on board in 1961 after hosting *Romper Room* programs in other cities.

Throughout most of the 1960s, the WHAS news was anchored by Fred Wiche. He would later be the station's longtime farm director and "weekend gardener," offering gardening news and advice in daily reports and on phone-in shows.

WHAS news anchor Fred Wiche works in the station newsroom in the mid-1960s.

Speculation was one of the last locally produced game shows. Emceed by Bob Kay and based on the stock market, it ran on WAVE from 1965 to 1966.

A somber Bill Gladden shows off his weathercast sponsor at WAVE in the early 1960s. In 1971, Gladden would leave WAVE and go to work for that sponsor, Greater Louisville First Federal Savings and Loan.

In the 1960s, thanks to videotape, *Hayloft Hoedown* could move out of the studio and feature song numbers filmed outdoors at different locations. In this 1965 shot, producer-director Bob Pilkington works with Jo Ann Hale on "Come Fly With Me" at Bowman Field.

Above is a view from the control booth at WHAS during a newscast in the mid-1960s. Anchor Fred Wiche is in the center-right monitor.

WAVE newsman John Lucy operates a film camera on a news shoot in 1967. Putting a story together was a time-consuming process involving shooting, developing, and editing film.

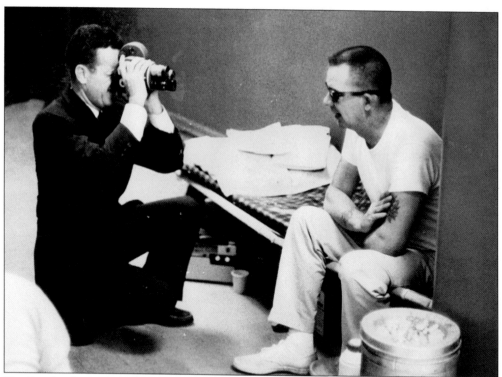

Ken Rowland was one of WLKY's first news anchors and reported many stories himself as well. Here he photographs an interview subject for a 1966 story.

Ken Rowland does the WLKY news in 1966.

After being named America's Junior Miss in 1963, Diane Sawyer went to college and then returned to her hometown. She joined WLKY in 1967, beginning as the station's "weather maid" but soon moving to a reporter position.

The advent of videotape in the early 1960s meant that live shows would soon be a thing of the past. Here at the WHAS studio, a technician oversees the taping of *Popeye's Adventures* with Randy Atcher in 1962.

T-Bar-V Ranch was still going strong in the 1960s with a new set and a new character, seen between Cactus and Randy—a puppet created by artist Allen Blankenbaker called WHAS Gnu (the WHAS stood for William Harris Allen Smith). There was also a donation box decorated with WHAS mascot FISBIE to accept donations for the "Crusade for Children."

It's an exciting romp of fun, fancy and fact, when Randy Atcher and Cactus Tom Brooks open the window of the world on T-BAR-V.

Blending live entertainment with fantastic cartoons and educational features, T-BAR-V exposes children to information in a palatable and interesting form.

Outstanding in these forms is the unsurpassed Encyclopaedia Britannica film library with hundreds of features involving animals, our world, fairy tales, children of other lands, safety, etc. The entire library has been designed to develop healthy minds and keen intellects.

Mrs. Barbara Miller and Story Time

Selected children's stories are read by Mrs. Barbara Miller, head of the Children's Department of the Louisville Free Public Library. This section brings fascinating stories to the youngsters and encourages them to join in the fun of reading.

The "Helpful Homer" slide series, produced by Encyclopedia Britannica, teaches youngsters about everything from clouds and seagulls to zebras and agriculture. Cactus reads the narration.

Locally Animated Songs and Stories

Former Walt Disney artist Allen Blankenbaker has produced a series of video taped animations to illustrate recorded songs and stories for children.

Guests demonstrate activities such as kitemaking and sand casting. **Specialty acts** with scouts, little league players, ice show performers and circus clowns keep the entertainment varied. A **"How To"** series teaches youngsters to make paper hats, boats, airplanes, etc., using paper, spools, beads and other easy-to-obtain items.

Britannica Films and Live Animals

More than simple entertainment, T-BAR-V is an exciting picture of the world for youngsters in the WHAS-TV area.

"Helpful Homer" and vivid lessons

WHAS-TV *Fisbie* **11**   "Foremost In Service—Best In Entertainment"

A 1962 sales sheet for *T-Bar-V Ranch* includes information on Barbara Miller, then head of the children's department at the Louisville Free Public Library and a regular visitor to the show.

This scene from a WAVE Christmas special in the mid-1960s includes a choral group.

Randy Atcher and Judy Marshall pose on the set of *Hayloft Hoedown* in the mid-1960s.

In 1961, the *Cartoon Circus* afternoon show on WHAS welcomed a new cast member—a parrot named Lolita who "spoke" in a sped-up voice and commented on everything from the space race to Mickey Mantle. Lolita is remembered as a disagreeable, finger-biting costar.

Off camera, "Cactus" Tom Brooks was a natty dresser who loved tailored suits. So he willingly participated in a 1963 stunt for Clean-Up Week where he wore a tuxedo and a mask to see if anyone would recognize him. Here he is outside the WHAS studio with Janet Risley, age 7, and Barbara Risley, 3. (Barbara Risley Dempsey.)

Three

1971–1985
TALK SHOWS AND LIVE SHOTS

In 1971, longtime local favorite *T-Bar-V Ranch* had bitten the dust, and *Hayloft Hoedown* was also on the way out. But local programming got a boost when independent station WDRB went on the air. Its schedule included reruns of *The Munsters* and *The Patty Duke Show* and first-run syndicated shows like *Rocket Robin Hood* and *Ultraman*, but the station also featured local programming, including *Presto the Clown* and the Saturday night *Fright Night* movies introduced by the Friendly Fearmonger. Local talk shows kept rolling along. Over at WAVE, *The Morning Show* began looking at controversial topics. At WHAS, *Omelet* teamed Milton Metz and Faith Lyles, and the show would run from 1971 until 1976.

Competition between local newscasts also began to heat up. At WHAS, station managers were tired of playing perennial second fiddle to WAVE's Livingston Gilbert, so in 1970 the station hired Mort Crim from ABC radio in New York and Ken Rowland from WLKY. The result was a gain in ratings, but Gilbert and WAVE were still on top. Meanwhile, WAVE did not stand still. In 1976, the station hired its first female coanchor, Melissa Forsythe, and was the first to use the new minicam for on-the-spot stories.

Meanwhile, over at WLKY, the star on the rise was Ange Humphrey. She went from doing commercials and hosting the *Dialing for Dollars* afternoon movie to doing the weather on the nightly newscast. Her attractiveness and down-home manner quickly won a strong following. Soon Humphrey was branching into a short-lived country music career and appearing on Tom Synder's *Tomorrow* show on NBC.

By the end of the 1970s, with the help of coanchors Jim Mitchell and Kirstie Wilde, WHAS was finally gaining on WAVE, and Livingston Gilbert was looking at retiring. WAVE brought in Dan Cullen as a coanchor with Melissa Forsythe, but viewers did not take to the combination. Cullen left and Forsythe was taken off the air, so she signed a contract with WHAS. WAVE sued to keep her from leaving, but in the end Forsythe joined WHAS.

Meteorologist Tom Wills joined WAVE in 1969 and would stay with the station for 40 years. As one of the area's first TV meteorologists, he had an authoritative, non-flashy manner that came to symbolize weather credibility for a generation of viewers. Wills is seen in a 1970 photograph.

WHAS personality Jeff Douglas hosted *Now* in the early 1970s. Douglas was a popular deejay on WHAS radio who also did duty on television. *Now* was an updated version of *Hi Varieties*, with the focus shifted to current events and group discussions. A talented man with an understated wit, Douglas committed suicide in 1976. (Eileen Douglas Israel.)

In 1970, WHAS hired Mort Crim (above, left) and Ken Rowland to coanchor the news. Crim came from ABC radio in New York, and Rowland came over from WLKY. Rowland would stay with WHAS until 1977, when he returned to WLKY.

WHAS coanchor Mort Crim is shown in a photograph at left from the early 1970s.

NEIL BOGGS

KEN ROWLAND

TV 11 NEWS
A LOOK AT KENTUCKIANA
NEWS THAT'S TO THE POINT, FOR PEOPLE LIKE YOU.

11 whas·tv

In 1972, Mort Crim left WHAS, and Ken Rowland was teamed with Neil Boggs to deliver the news. The low-key Boggs left some viewers cold and departed after a year or so, and WHAS news continued its pursuit of WAVE, the ratings leader.

When WDRB signed on the air in 1971, the last local kids show had just bitten the dust. Then came Presto the magic clown, played by magician Bill Dopp. His daily show, which included sleight of hand and tricks like "Sands of the Desert," was a big success for WDRB and ran until 1976. Dopp died in 1994 at the age of 79.

The host of *Fright Night* was the Friendly Fearmonger, played by local actor Charles Kissinger, shown here in a 1973 ad. The act and its special effects were minimal. Kissinger told creaky monster jokes in a dark studio illuminated by only a red light just under his chin.

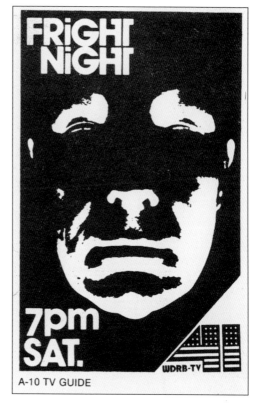

The Friendly Fearmonger was, in real life, Charles Kissinger. He also appeared in several cheesy monster movies filmed in Louisville. Kissinger died in 1991 at age 66.

The WAVE news old guard is shown here around 1972. Even after more than 20 years, WAVE was still the ratings leader, thanks to Livingston Gilbert, Ed Kallay, Tom Wills, and Rodney Ford.

In the early 1970s, Milton Metz stepped down as WHAS weatherman to cohost *Omelet*, a talk-variety series with Faith Lyles. Metz also hosted a call-in show on WHAS radio that ran from 1959 until 1993.

Omelet, the WHAS entry into the morning talk show sweepstakes, ran from 1971 to 1976. Cohosts were Milton Metz and Faith Lyles, seen in this 1975 ad.

In 1972, Bob Terry hosted WLKY's afternoon movie. He was also the station's weatherman and in the early 1970s hosted a morning kids' show, *Bob Terry and his Pirates*.

Julie Shaw and Ryan Halloran, hosts of *The Morning Show* on WAVE-TV, pose on the set in the early 1970s. Halloran and Shaw would leave the show in the mid-1970s, succeeded by younger cohosts. The show itself would run until 1980.

To celebrate WAVE-TV's 25th anniversary in 1973, Ed Kallay and Julie Shaw recreated their *Magic Forest* roles as Uncle Ed and Miss Julie the Story Lady.

Pee Wee King and his band appeared on WAVE's 25th anniversary show in 1973.

News anchor Lee Hunter (left) was next up in the battle between WHAS and WAVE news. He joined Ken Rowland at WHAS in 1974 and left in 1976.

Ed Kallay (below, right), attending a 1973 banquet, chats with a boyhood friend, Olympic great Jesse Owens. At this same event, honoree Muhammad Ali turned the spotlight on Kallay. "I was just a poor boy here in Louisville who rode his bicycle to the gym. The chamber of commerce was not with me then. But Ed Kallay was."

Weekdays at 3:30 PM
DIALING FOR DOLLARS MOVIE
with Ange Humphrey
32 WLKY·TV
louisville, ky.

Before becoming the weathercaster at WLKY, Ange Humphrey did local commercials and, in 1973, hosted the station's afternoon movie.

WEEKEND NEWS TEAM
6 & 11 Sat., 11:15 Sun.
Monica Kaufman, Byron Crawford
11 whas tv
LOUISVILLE, KY.

Monica Kaufman and Byron Crawford anchored the WHAS weekend newscast in 1975. Crawford went on to report "On the Road" segments for WHAS and then became a longtime columnist for *The Courier-Journal.* Kaufman would become a well-known anchor in Atlanta. Her moment of infamy on WHAS came when she failed to realize she was still on the air and said "shit" after flubbing the introduction to a story. (She hastily apologized.)

WLKY weathercaster Ange Humphrey poses in a farm field for a promotional spot in the mid-1970s. (WLKY.)

Ange Humphrey's looks and plain-spoken style made her a local media superstar in the mid-1970s.

A "family" photograph of WAVE personalities from the mid-1970s included, from left to right, (first row) sportscaster Bob Domine, newscaster Mary Ann Childers, meteorologist Tom Wills, and longtime news anchor Livingston Gilbert; (second row) newscaster David Robinson, newscaster Dale Solly, sportscaster Randy Waters, newscaster Bob Kay, and coanchor Melissa Forsythe; (third row) sports director Ed Kallay, weatherman Norm Lewis, and newscaster Kathleen Partlow. (WAVE.)

Dave Nakdimen was with WAVE news from 1961 to 1997 as a government and political reporter. (WAVE.)

A photograph from the mid-1970s shows Fred Wiche (left) and Barney Arnold, the man Wiche would succeed as WHAS farm director.

THE WEEKEND GARDENER

7:30PM FRED WICHE visits Farmington, one of Louisville's original homes and compares foliage decorating, now...plus tips on designing Christmas wreaths.

MONDAY

WHAS 11 ©

After almost 20 years as a reporter and anchor for WHAS news, Fred Wiche flawlessly executed a career change in the mid-1970s. He parlayed his love for gardening into a role as the station's horticultural expert, later becoming farm director as well. Wiche dispensed gardening advice on WHAS radio and television until just before his death in 1998.

On the eve of the nation's bicentennial in 1976, WDRB asked viewers to believe that the original patriots enjoyed watching *Ultraman* every afternoon.

At WAVE, sportscaster Bob Domine was as well known for trying outrageous stunts in his "Domine Does It" segments as he was for delivering scores and game news. Domine joined WAVE in 1973 and, except for a short stint at a station in Philadelphia, was there until his retirement in 2010.

In 1976, the minicam changed local TV news forever. A premium was placed on live reports from the scene—even if the story had happened hours ago. WAVE was the first station in town to get a minicam.

In 1976, Melissa Forsythe became Louisville's first woman television news anchor, assuming the chair opposite Livingston Gilbert on WAVE news. A native of Southern Indiana, she had risen through the station ranks, including a stint as a photographer.

Rick Moore was the anchor of WLKY news in the mid-1970s.

WAVE anchor Livingston Gilbert is on the set in the late 1970s.

On April 30, 1977, after making a personal appearance in a parade, Ed Kallay stopped at a supermarket to pick up some milk on the way home. There he had a heart attack, collapsed, and died. Kallay was 59.

In 1977, Jim Mitchell came on board at WHAS and would stay with the station until 1988. During his tenure, WHAS news finally launched a serious challenge to WAVE.

In 1978, preceded by hype that included billboards, TV spots, and newspaper articles, Kirstie Wilde came to WHAS news, joining coanchor Jim Mitchell (left), meteorologist Chuck Taylor, and sports director Dave Conrad (far right). With her arrival, WHAS's news ratings continued to rise.

As Livingston Gilbert prepared to retire, Dan Cullen (left) was brought in to coanchor WAVE news with Melissa Forsythe.

Viewers did not take to Dan Cullen as WAVE anchor, and he left within a year.

LOUISVILLE TONIGHT

with
Tom & Ange
Weeknights 7pm

premiering Sept 17
WHAS 11

In 1979, *Louisville Tonight* joined the parade of local magazine shows, highlighting feature stories about the surrounding area. Chosen as the first cohosts were WHAS reporter Tom Van Howe and former WLKY weathercaster Ange Humphrey. With several sets of hosts, *Louisville Tonight* would have a 20-year run.

Cawood Ledford broadcast University of Kentucky basketball and football games from 1953 until 1992 and was with WHAS from 1956 to 1979. This ad is from 1979.

The WHAS Crusade for Children, celebrating its 25th year in 1979, welcomed celebrities Tim Reid (Venus Flytrap on *WKRP in Cincinnati*), Bobby Rydell, Della Reese, and Bob Keeshan, better known as Captain Kangaroo, shown below from left to right.

99

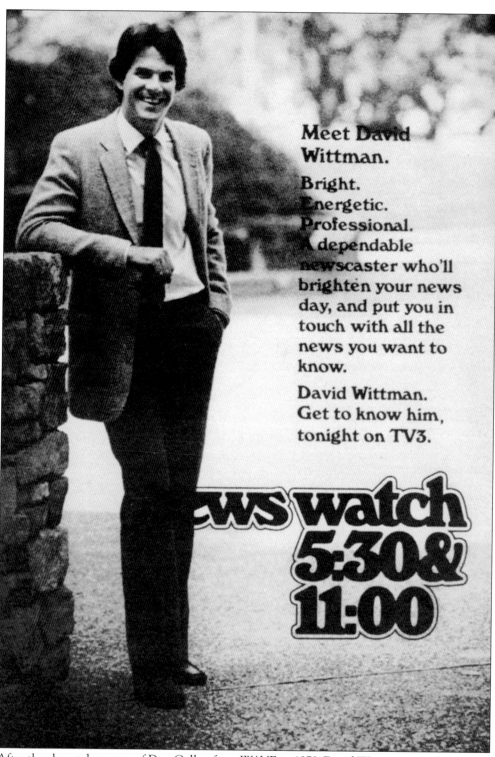

Meet David
Wittman.

Bright.
Energetic.
Professional.
A dependable
newscaster who'll
brighten your news
day, and put you in
touch with all the
news you want to
know.

David Wittman.
Get to know him,
tonight on TV3.

ews watch
5:30&
11:00

After the abrupt departure of Dan Cullen from WAVE in 1979, David Wittman joined the news team as coanchor.

By 1980, Livingston Gilbert (center) was still at WAVE, but he had two new coanchors—David Wittman and Ann Wade. The station's advertising was imploring viewers to "come home." By the end of that year, Gilbert had retired.

WLKY played up Ken Rowland's credibility in a 1980 advertisement. In 2005, Diane Sawyer paid tribute to Rowland, her first boss in TV news: "The Nobel Prize would be nothing compared to an e-mail from Ken that says 'That was good.' "

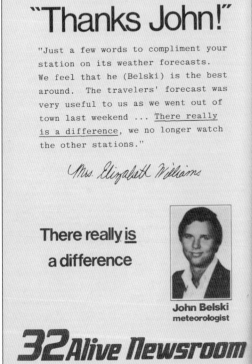

In 1980, WLKY meteorologist John Belski single-handedly saved one family's vacation from a weather-related disaster.

Jackie Hays is shown just after her arrival at WHAS in 1980. (WHAS.)

In 1982, Kirstie Wilde (right) decided to leave WHAS and return to her native California. Jackie Hays (center) stepped in as the new WHAS news coanchor with Jim Mitchell. (WHAS.)

WHAS-TV celebrated its 30th anniversary in 1980 by welcoming back two performers it had dumped in 1971—Randy Atcher and "Cactus" Tom Brooks. (Author collection.)

Liz Everman began her career at WLKY in 1980. She retired as coanchor in 2005 but continues to do her "Wednesday's Child" segments, which have helped almost 3,000 special-needs children find homes. (WLKY.)

When WHAS coanchor Jackie Hays left for a Philadelphia station in 1985, former WAVE anchor Melissa Forsythe stepped in.

In July 1983, WDRB tried something totally unique. The station ran dozens of shows from television's golden age in primetime for an entire month. The programs were hosted by promotion director Ray Foushee.

Four

1985–TODAY
ALL IN THE NAME OF NEWS

By the mid-1980s, the Louisville television landscape was much different than it had been in the 1960s. There were no more kiddie shows, no more talent shows, no more music shows. There was only news: early morning news, noontime news, early afternoon news, late afternoon news, late-night news, weekend news. Ratings battles between local stations raged on, but the winning margins were smaller and smaller, and the winning stations changed on a regular basis.

In 1987, WDRB joined the new Fox network. The station began its own local late-night newscast in 1990; the first anchors were Jim Mitchell, formerly of WHAS, and Lauretta Harris, formerly of WAVE.

In 1990, WHAS announced that after 40 years, it was switching its network affiliation to ABC from CBS. This left CBS and WLKY in the lurch, so they teamed up.

By the end of the 1990s, WDRB was also programming early morning and afternoon news. In 2006, WHAS began producing a nightly 10:00 p.m. newscast airing on local CW affiliate WBKI.

And by the late 2000s, each local station featured extensive Web sites with blogs written by reporters and meteorologists.

On his last night as WLKY coanchor in 1987, Ken Rowland gets ready to go on the air. (WLKY.)

Upon completion of his final WLKY newscast, Ken Rowland gets an affectionate punch from coanchor Liz Everman. (WLKY.)

WHAS coanchors Jim Mitchell and Melissa Forsythe are on the set in the late 1980s.

Melissa Swan began her career at WLKY before moving to WHAS in 1985. Soon after that, she became coanchor of the WHAS evening news. (WHAS.)

Gary Roedemeier joined WHAS as a coanchor in 1984. (WHAS.)

Doug Proffitt began at WHAS as an intern in 1978 and became a reporter in 1987. Now he is coanchor of the station's nightly newscasts. (WHAS.)

Rachel Platt joined WHAS in 1989 and is the longtime coanchor of *Good Morning Kentuckiana*. (WHAS.)

In 1988, WLKY meteorologist John Belski moved to WAVE. He retired in 2010. (WAVE.)

Ferrell Wellman was the Frankfort reporter for WAVE for over 15 years. He now hosts *Comment on Kentucky* on Kentucky Educational Television. (WAVE.)

From still photographs to celluloid film to digital editing, local TV news production has changed immensely over the last 60 years. Below WAVE sportscaster Bob Domine gets his report ready. (WAVE.)

After several years as an anchor in Philadelphia, Jackie Hays decided to return to Louisville. She joined WAVE news in 1988. (WAVE.)

Melanie Bullard was a WAVE reporter and anchor from 1986 to 1995. (Melanie Bullard.)

In 1990, WDRB launched its first newscast. *The News at 10* featured coanchors Jim Mitchell and Lauretta Harris. (WDRB.)

WAVE anchors Jackie Hays (seated), Don Schroeder, and Melanie Bullard pose for this image in the early 1990s. (Melanie Bullard.)

WAVE anchors Jackie Hays, Melanie Bullard (seated), and Jim Mitchell are photographed in the mid-1990s. Both Mitchell and Hays had worked together at WHAS. (Melanie Bullard.)

In the mid-1990s, Bob Sokoler and Kirby Adams hosted *Louisville Tonight*. (WHAS.)

Richard Simmons's visit to *Louisville Tonight* in 1994 prompts a quick shushing from cohost Kirby Adams. (WHAS.)

The *Louisville Tonight Live* on-air staff in late 1993 are, from left to right, cohost Kirby Adams, reporter Lee Thomas, reporter Jayne McClew, cohost John O'Connor, and reporter Roseann Rogers. (Jayne McClew.)

Vince Staten
Jayne McClew

Louisville
TONIGHT
Live

KENTUCKIANA'S
NEWS CHANNEL
WHAS⑪

In the mid-1990s, Jayne McClew and Vince Staten appeared as the "Movie Police" on *Louisville Tonight Live.* (Jayne McClew.)

WAVE newscaster Hugh Finn poses for a publicity shot. A serious injury in a 1995 car accident left him in a persistent vegetative state. His wife, Michelle, fought to have his feeding tube removed, which resulted in a family dispute lasting three years. A court ruling finally allowed for the removal of the tube in late 1998, and Finn died a few days later. (WAVE.)

Melissa Forsythe was an anchor with WHAS until 1991. (WHAS.)

In the 1980s and 1990s, WLKY sports director Fred Cowgill was always willing to take on stunts as part of his sportscast. In this 1994 photograph, he tries his hand at cattle roping while then–University of Louisville basketball coach Denny Crum watches in the background. (David R. Lutman.)

This photograph shows WLKY sports director Fred Cowgill. (WLKY.)

Reed Yadon, longtime newsman and sidekick to deejay Bill Bailey on WAKY radio, joined WLKY as a weatherman in 1979 and stayed at the station until 1994. He then joined WHAS as a weathercaster and helicopter pilot in the late 1990s. (WHAS.)

By 1993, Fred Wiche's "Weekend Gardener" empire included regular spots on WHAS radio and TV, books, and a syndicated weekly newspaper column. (WHAS.)

In the mid-1980s, Christy Callahan was given the star treatment before becoming a coanchor with Don Schroeder on WAVE. Commercials told viewers, "Christy Callahan—she fits right in!"

John Belski   **Rick Little**   **Liz Everman**   **Fred Cowgill**

Weeknights 6 & 11pm

CHANNEL 32 NEWS

Ken Rowland stepped down as WLKY coanchor in 1987. He had been with the station off and on since just after it went on the air in 1962. He was replaced by coanchor Rick Little.

Barry Bernson has worked in the news departments of three Louisville stations—WAVE, WHAS, and WDRB—and spent some time at a Chicago station. Long a local favorite who excels at offbeat features, Bernson now cohosts WDRB's *Fox in the Morning* with Candyce Clifft. (WDRB.)

Elizabeth Woolsey and David Scott are the current anchors of WDRB's *Fox News at 10*. (WDRB.)

In March 2009, longtime WHAS reporter Chuck Olmstead died of a ruptured brain aneurysm. He had been with the station for 34 years. WHAS and WLKY hosted telethons in his memory to benefit the Chuck Olmstead Memorial Fund, which would be used to purchase portable MRI units for free screenings. (WHAS.)

There are two constants in Louisville TV history—the Kentucky Derby and the "WHAS Crusade for Children." Here are WAVE anchors Scott Reynolds and Jackie Hays at Churchill Downs. (WAVE.)

Thanks to satellite transmission, on-the-spot reports are more common than ever. Here WHAS newsman Mark Hebert reports live from a wintry fire scene in Louisville's Highlands neighborhood. (Steve York.)

In August 2009, WAVE meteorologist Tom Wills (left) announced he would be retiring after 40 years at the station. Here he poses with WAVE meteorologist Kevin Harned. (Steve York and WAVE.)

On his last day as WAVE meteorologist, Tom Wills gives the forecast and chats with anchor Lori Lyle. (Steve York and WAVE.)

Late in 2009, WAVE anchor Jackie Hays announced she would be leaving WAVE after 21 years. Moments before her final newscast, Hays and coanchor Scott Reynolds prepared to go on the air. (Steve York and WAVE.)

On the night of her last newscast, Jackie Hays (seated third from left) poses with, seated from left to right, anchor Dawne Gee, anchor Scott Reynolds, and meteorologist John Belski. WAVE sports director Kent Taylor stands at the back. (Steve York and WAVE.)

DISCOVER THOUSANDS OF LOCAL HISTORY BOOKS FEATURING MILLIONS OF VINTAGE IMAGES

Arcadia Publishing, the leading local history publisher in the United States, is committed to making history accessible and meaningful through publishing books that celebrate and preserve the heritage of America's people and places.

Find more books like this at
www.arcadiapublishing.com

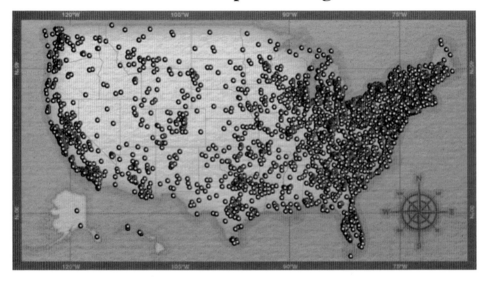

Search for your hometown history, your old stomping grounds, and even your favorite sports team.